ULTIMATE BOOK TO DRONES

I0490563

Self practical manual for drone operations and repairs guide

Karen D. Murphy

Table of Contents

CHAPTER ONE

INTRODUCTION TO DRONES

Drone has different types, but all it type serve one purpose is created to general information's to the owners. A Remotely Piloted Aircraft (RPA) is a type of airplane which has no onboard crew or passengers.

An RPA is profitable of controlled, sustained stage flight and is powered by way of capability of a jet, reciprocating, or electric powered engine. Crucially the RPA's important characteristic is to be the issuer of a payload. Payloads can vary considerably from a producer built-in digital camera for photos or videography to pretty specialized payloads such as LIDAR or Gas Sampling. There are many precise payloads used for many first-rate missions. It is the functionality to unexpectedly and exactly function the drone and payload in the excellent location to capture the relevant records that affords blessings in correctly and safety.

CHAPTER TWO

TYPES OF DRONE

> ## CLASSIFICATIONS OF DRONES

✓ **Guide In Describing Fixed-Wing Drones**

Fixed-wing drones are these profitable of harnessing air and producing forces that allow them to stay in the air by using taking gain of their aerodynamics.

They are same in design or aesthetics to radio-controlled aircraft and are regularly used to map large areas due to their tremendous autonomy. They take obtain of their aerodynamics and layout to maintain them afloat, which practicable they have a longer persistence and flight speed. The downside of fixed-wing drones is that they have a tendency to be greater luxurious in distinction to multi-rotor drones. They require a large, clear region to take off and land, simply like airplanes. Some massive fashions moreover require specialized flooring equipment to help them take off and land. In addition, fixed-wing drones can totally fly forward, so they do no longer furnish the equal maneuverability as multi-rotor drones.

✓ Guide In Describing Multi-Rotor Drones

Multi-rotor drones, moreover stated as rotary-wing, are the most considerably used form of drones for entertainment and professional use.

Their small dimension and pinnacle notch control make multi-rotor drones the nice desire for aerial photography. Offering high-quality versatility, they allow the set up of all sorts of cameras to characteristic special tasks. They are drones that can hover without difficulty and take off vertically, such as higher flexibility as well.

However, the biggest downside of multi-rotor drones is usually related to the flight autonomy they offer. Adding greater rotors makes the drone larger difficult to control. All these transferring elements moreover devour increased energy, draining the battery faster. Most multi-rotor drones have a flight time of a great deal much less than an hour. Function positive obligations that take a prolonged time, we have to have infinite batteries to alternative them. This moreover capability extra costs.

✓ Guide In Describing Single-Rotor Helicopter Drones

Powerful and durable, single-rotor drones appear to be same in constructing and structure to authentic helicopters, with completely one rotor to provide power, plus a tail to manipulate route and stability. Combining the advantages of tiny multi-rotor drones and single-rotor drones, they are greater ideal to increase massive payloads and fly increased efficiently than multi-rotors. Single-rotor devices usually use fuel engines as a substitute than batteries, which radically will enlarge their flight time.

However, these drones have a tendency to be massive and increased tricky than distinct kinds of UAVs. This functionality they are higher high priced and larger difficult to operate, and their massive blades can make them larger dangerous.

✓ **Guide In Describing Fixed-Wing Hybrid VTOL Drones**

As the existing day drone technological understanding to be introduced, fixed-wing hybrid VTOL drones refer to fixed-wing airplane that have been modified to take off and land vertically. They combine the long-range and flight time of fixed-wing UAVs with the vertical takeoff performance of rotary-wing devices, casting off the drawbacks of fixed-wing UAVs that require large areas for takeoff and landing. They are designed for mapping, power line inspection, surveillance, agriculture, and rescue operations.

> **<u>Engine Part Of Drones</u>**
✓ **Drones Frame**

It ought to have enough electrical energy to hold the propeller momentum and more weight for motors and cameras, Sturdy and a whole lot much less aerodynamic resistance

✓ Drone Propellers

The speed and load lifting functionality of a drone depends upon on shape, size, and vary of propellers, The prolonged propellers create massive thrust to elevate heavy lots at a low speed (RPM) and a whole lot much less sensitive to change the speed of rotation, Short propellers carry fewer loads. They alternate rotation speeds rapidly and require an immoderate tempo for greater thrust.

✓ **Drone Motor**

Both motors brushless and brushed variety can be used for drones, A brushed motor is tons much less high-priced and really helpful for small-sized drones, Brushless sort motors are high quality and electrical energy very efficient. But they want Electronic Speed Controller (ESC) to manipulate their speed. These brushless motors are widely used for racing freestyle drones, website online visitor's surveys and aerial photographs drones.

✓ Drone Electronic Speed Controller

ESC is used to be part of the battery to the electric powered motor for the electrical energy supply, It converts the signal from the flight controller to the revolution per minted (RPM) of motor, ESC is furnished to each y motor of the drone

✓ Drone Flight Controller

It is the laptop computer processor which manages balance and telecommunication controls the use of splendid transmitter, Sensors are positioned in this unit for the accelerometer, barometer, magnetometer, gyrometer and GPS, The distance measurement can be carried out with the useful resource of an ultrasound sensor, Radio Transmitter sends the radio signal to ESC to pilot to control motor speed.

✓ Drone Radio Receiver

Received the signal from the pilot, this system is related to the quadcopter.

✓ Drone Battery

High-power capacity, Lithium Polymer LiPo is used for most drones. The battery can have 3S 3 cells or 4S 4 cells.

✓ **Drone Landing Gear**

This is a form supposed for safely landing the drone. However, it can be exempted due to the reality an expert purchaser is profitable of balancing the motors tempo for included landing in emergencies. There are two most essential kinds of landing gear. One is steady landing equipment and the special is retractable landing gear.

CHAPTER THREE

COMMON ISSUES OF DRONES

❖ **When Drone Lack Syncing**

Often quadcopters no longer connecting to the controller due to the reality of insufficiently charged batteries, and an unpaired controller to the quadcopter.

✓ *What to do to avert lack syncing*

Charge the quadcopter and controller truly or to an adequate diploma beforehand than a flight.

If the quad is blinking pair the controller to the quadcopter. Push the left throttle up and wait until the controller beep and the quadcopter's LEDs are secure Else blinking Turn on the drone then hold the desirable trim on the far off until the quad provide up blinking it can also moreover take about 1 minute to reset the link.

❖ Instance Of Drone Digicam Camera Failure

In most cases, the drone digital camera now not working or now no longer showing the video feed due to the truth of the out of date app glitches, circuit issues, and crashes.

✓ *How to avert it*

Clear cache and app information of the app. uninstall the contemporary app on your phone. Uninstall the app on your Smartphone and set up the modern up to date mannequin of the app. Repair the drone. If you crashed recently, the drone circuit or the digital can also have been damaged.

❖ **Drone   Unable   To Connect Wifi**

Usually, drone won't be a part of to wifi due to the reality of wifi glitches, or mobile utility issues. Also if you have paired the drone to each different computer recently, it moreover would motive this issue.

✓ *Steps to solve and fix it*

Restart the mobile machine and strive again. Turn on aircraft mode and flip off it. When you flip on plane mode, it would definitely shuts down the neighborhood on your device. Then when you flip on it as soon as greater it would strive to examine the new devices and reconnect them. Enable

Networking and Wireless option. Go to Settings, Privacy, Location Services, and computing device choices on your device. Then permit the networking and wi-fi option. If it's already enabled, disable it and enable it again. Try with any different device.

❖ **Drone App Unable To Work**

In most cases, the app now no longer working and responding due to the truth of an old school app, incompatible mobile device, or placing in an Apk file with glitches.

✓ *Steps to fix it*

Check your Smartphone is nicely matched with the app. Clear the cache data and app records of the app. Go to Settings > Apps > app and clear cache data and app data. Uninstall the app and set up the current mannequin of the app. If the nowadays established app or its settings may additionally have bugs, consequently first uninstall it and set up the app as soon as greater as an choice of absolutely updating it.

❖ Unable To Fly Drone

If the quadcopter is no longer taking off even though it's the propellers are spinning, you ought to have hooked up them in the flawed pattern.

✓ How to fix and service

Check whether or not or no longer are propellers hooked up correctly. If your drone won't take off even though propellers are spinning you would perhaps have hooked up them in the mistaken pattern. And Strive putting propellers in distinct pattern. Inspect the drone model and set up propellers in accordance to the model.

❖ **Drone   Lack   Of Calibrating**

Most situations the quadcopter calibrate manner continues failing due to the truth of performing the calibration manner on a titled surface.

✓ *Easy ways to fix it*

Calibrate the drone on a flat surface. When you are calibrating the drone you truly have to use a definitely flat surface. If you are making an attempt to calibrate the drone on uneven flooring it will lead to calibration failure or inaccurate calibration results.

❖ Drone Unable To Charge Properly

Usually, drone refuses to price due to the reality of damaged battery cells, inaccurate or broken charger, or low amperage and voltage.

✓ *What to do*

Check whether or not or now not the charger is Faulty or broken. Inspect whether or not or now not the charger output amperage and voltage fulfill the required inputs for the battery. Plug the charger into a power outlet with the required voltage and current. The battery is swollen or damaged, alternative the battery. Most of the time lithium polymer batteries have a tendency to swell as quickly as they are used for a prolonged time. So it's endorsed to alternative the battery in these sorts of situations.

❖ Drone Constant Low Battery

Usually, the drone no longer working or no longer turning on due to the reality of low battery power, damaged battery cells, or circuit failures.

✓ *How to fix it*

Charge the drone battery to enough diplomas formerly than turning on. Replace the battery. If you are noticing swelling or some damage to the battery change the battery. Repair the drone. If you crashed recently, the drone circuit has been damaged. So you would have to take aid from experts to repair the drone.

❖ When Drone Propeller Is Misalign

Sometimes, quadcopter's some propellers are now no longer working due to the reality of misalignment between motor gear and propeller gear or loss of one of them. Tangled motors, battery issues or damaged wires moreover might also desire to this range of situation.

✓ *What to do*

Check whether or not or now not the motor equipment is missing. Uncover the drone safely and take appear at the tools place of the malfunctioning propeller. If you have a look at one of the gear is missing, check internal the drone for that popped gear.

If you positioned it, install it gently and align it with exclusive gears. If you misplaced equipment you would have to buy new gear. Usually, they are now no longer expensive. Look for damaged, shorted wires. If you stumble upon ripped or shorted wires, fix them. Replace the motor. If the motor is clear and now no longer spinning, the motor may additionally have been damaged. So you have to alternate them.

❖ Lack Of Steady Stable Drone

The main motive for the instability of the quadcopter is miscalibration. Besides that, mal-functioning motors, short wires, and distorted propellers cause the issue.

✓ *Comprehensive steps on fixing it*

Calibrate quadcopter. Turn on the quadcopter, pair it to the controller, and go away it on a flat surface. Then pull every controller sticks to the appropriate bottom nook to calibrate the drone. Replace propellers with a new set of propellers if some are distorted. Inspect tools laptop of malfunctioning propellers.

If the quadcopter is on the other hand unstable after calibration, then there may additionally be a misalignment in the gear system. Uncover the quadcopter and take seem at whether or not or no longer all the gears associated with the motor are flawlessly aligned.

Replace malfunctioning motors. If the quadcopter is nonetheless flying sideways or unstable, some motors would maybe malfunction. Try changing motors spinning low.

Drones comes in different sizes and it works base on it size, types and component part

Drone of very small size and component part

The dimension of these drones can be as small as an insect and barring issues swimsuit in the palm of your hand, with dimensions between 1 and 50 cm. Because of their small dimension and wing design, drones can fly in very restrained areas and are now no longer barring issues detected and are normally used by way of the usage of spies to accumulate documents on human beings and objects.

Drone of small size and component part

Small drones they are barely giant than micro drones, between 50 cm and two meters. The wings of these drones are commonly regular and can be without problems lifted with the resource of the arm and thrown into the air. They can be used for indoor tools inspections prop guards, cages; then again they are greater normally used exterior for exercise and photography, such as web site visitor's management.

Drone medium size and component part

Medium-sized drones are giant and heavier than smaller drones, with dimensions exceeding two meters and weighing up to 200 kilograms. They take two human beings to raise them, which are most often used for specialist functions and novice photography.

Drone large size and component part

Large drones are associated in dimension to smaller aircraft and are used usually for navy features such as surveillance and strategy. With the most optimum technology, they are altering fighter jets, allowing for quickly enemy detection and combat capabilities. They are moreover designed for civil functions such as drone deliveries or filmmaking.

www.ingramcontent.com/pod-product-compliance
Lightning Source LLC
Chambersburg PA
CBHW070801220526
45467CB00017B/791